ATE DUE

sun
power

sun power

Facts about Solar Energy

STEVE J. GADLER
WENDY WRISTON ADAMSON

LERNER PUBLICATIONS COMPANY ▪ MINNEAPOLIS

Acknowledgments

The illustrations are reproduced through the courtesy of: pp. 6, 26, 28, Hale Observatories, California Institute of Technology; p. 11, Gulf Oil Corporation; p. 13, Bureau of Mines, United States Department of the Interior; p. 15, E. I. Dupont De Nemours & Company; p. 17, Geological Survey, United States Department of the Interior; p. 19, Pacific Gas and Electric Company, San Francisco, California; p. 21, M. Brigaud — Sodel — E.D.F. for Documentation Francaise Phototheque, Paris; pp. 23, 91, Central Office of Information, London, for the British Consulate General; p. 30, Egyptian Museum, Cairo, Egypt; p. 33, Lennart Larsen for the National Museum of Denmark, Copenhagen; p. 34, British Crown Copyright, Central Office of Information, London, for the British Consulate General; p. 36, Museo Nacional de Antropologia, Mexico City; p. 39, Photo, Science Museum, London; p. 40, British Crown Copyright, Science Museum, London; pp. 42, 83, Centre National de la Recherche Scientifique, New York; pp. 47, 51, Honeywell Incorporated; pp. 48, 62, 68 (top and bottom), Energy Research and Development Administration; p. 54, Commonwealth Scientifić and Industrial Research Organization, East Melbourne, Australia; pp. 58, 79, Aden and Marjorie Meinel, University of Arizona; pp. 65, 66, Lockheed Missiles and Space Company, Incorporated; pp. 70, 72, 76, National Aeronautics and Space Administration; p. 71, TRW Defense and Space Systems Group; p. 74, Agricultural Engineering Department, University of Florida; pp. 85, 87, Owens-Illinois, Incorporated; p. 90, Mitsubishi Electric Corporation and Taisei Corporation for Sunshine Project, Japan; p. 94, United States Army White Sands Missile Range, New Mexico; p. 97, Bell Laboratories, New Jersey.

LIBRARY OF CONGRESS CATALOGING IN PUBLICATION DATA

Gadler, Steve J.
Sun power.

Includes index and glossary.
SUMMARY: Discusses the past, present, and future use of solar energy and compares this energy source to the more widely used fossil fuels and water power.

1. Solar energy—Juvenile literature. I. Adamson, Wendy Wriston, joint author. II. Title.

TJ810.G33   621.47   77-92290
ISBN 0-8225-0643-2

1980 REVISED EDITION

Manufactured in the United States of America. Published simultaneously in Canada by J. M. Dent & Sons (Canada) Ltd., Don Mills, Ontario.

International Standard Book Number: 0-8225-0643-2
Library of Congress Catalog Card Number: 77-92290

3 4 5 6 7 8 9 10 85 84 83 82 81 80

Contents

	Introduction	7
1	Sources of Energy in the Modern World	9
2	The Sun as a Source of Energy	25
3	Human Attitudes toward the Sun	31
4	Early Experiments in Using Solar Energy	37
5	Solar Energy Today	43
6	The Solar Future	57
7	Solar Energy: Pro and Con	77
8	Planning for the Solar Future	89
	Glossary	99
	Index	101

Introduction

Energy! It's a word that we see every day in the paper and hear on the evening news. Perhaps it makes us think of huge steam-powered electric plants, of runners in a cross-country race, of a locomotive chugging across the plains or a rocket shooting to the moon. It probably *doesn't* make us think of the sun, and yet this bright star, the center of our solar system, is the source of almost all the energy that has ever existed on the earth.

Today we are all aware that our world is experiencing an "energy crisis." Does this mean that there are almost no supplies of energy left? No, but it *does* mean that we are running short of the traditional sources of energy—oil, coal, and natural gas. The current energy crisis is a sign that we humans have not managed our supplies of energy well. We have insisted on driving large cars that consume huge quantities of gas. We have overheated our homes in the winter and have kept our air conditioners running all summer regardless of outside temperatures. We have left the lights and the television set on even when we were not in the room. By doing all these things, we have used so much fuel, especially oil and natural gas, that serious shortages have developed. Today we can even see the end of the supplies of some fuels.

Fortunately it is not too late. We will not run out of energy sources nor will the modern world come to a grinding halt—*if* we begin now to plan for the future. Our plans must include a reduction of energy consumption by all the world's inhabitants, particularly the people of the United States. They must also provide for the development of new sources of energy. In recent years, some scientists have turned to nuclear energy as the answer to the world's energy problems, but it has not proved to be the safe, nonpolluting, and economical source of power that we need. Is there such a source anywhere in our world? Many believe that the sun can provide the energy that is needed. That glowing ball in the sky worshipped by the people of ancient times could very well be the ultimate solution to the energy crisis of today.

1

Sources of Energy in the Modern World

Today the world gets its direct energy supply from a very few sources. The primary ones are the fossil fuels—petroleum (oil), coal, and natural gas. Other less important sources of energy include nuclear energy, geothermal power, hydroelectric power, tidal waters, and the wind.

FOSSIL FUELS

Fossil fuels are deposits in the earth that have been formed by the decomposition of the remains of prehistoric plants and animals. Since this organic matter originally derived its energy from the sun, the fossil fuels themselves can be said to be products of solar energy. Today, more than 95 percent of the world's energy comes from the fossil fuels.

Petroleum

Perhaps because petroleum can be turned into dozens of products and is relatively easy to transport, store, and use, this "black gold" has become the basic form of energy in most parts of the modern world. About 40 percent of the energy used in the United States is supplied by petroleum.

Unfortunately, this useful fossil fuel is not plentiful in all parts of the world. Existing supplies of petroleum in the United States are inadequate to meet the nation's increasing demands. Petroleum *is* abundant in the Middle East, and the oil-rich Arab countries supply large amounts to the United States and to other countries. But this fact introduces a new consideration into the energy picture— politics. In 1973, the Arab countries carried out an oil *embargo*, cutting off supplies to the United States and other western countries for political reasons. Politics also affected oil supplies in 1979, when a revolution in Iran resulted in reduced oil production and a sharp increase in prices. Such events have made it clear that even if supplies of petroleum exist in one country, they will not necessarily be made available to other countries. Or they may be made available at such high prices that petroleum will not be an economically realistic source of energy. And of course even the Middle Eastern supplies of oil will not last forever. Experts predict that within the next century the world's oil reserves will probably be exhausted.

In addition to problems of supply, there are other difficulties connected with the use of petroleum. The extraction and transportation of oil causes various kinds of pollution. Offshore drilling of oil wells has resulted in enormous amounts of oil pouring into and polluting the oceans. It is very difficult to clean up this spilled oil, and even more difficult to plug up a leaking well. In addition, the "supertankers" that carry oil across the oceans have

Offshore oil wells such as this one can cause special kinds of pollution problems.

accidently dumped thousands of gallons into the world's waterways, befouling beautiful coastlines, destroying millions of birds and fish, and seriously affecting tourist and fishing industries. Pipelines that carry oil overland have also had their problems; leaks from broken pipe have destroyed farmland and damaged the environment.

The pollution problems are far from over when the oil has reached its destination. The burning of oil for the production of electricity puts harmful pollutants such as sulfur dioxide into the air. And when petroleum is made into gasoline and other fuels used in transportation, it becomes the source of about half of all air pollution in the United States.

Coal

Coal, the most plentiful of the world's fossil fuels, supplies about 20 percent of the energy consumed in the United States. It is used primarily in industry and for the production of electricity. Unlike petroleum, coal is fairly abundant in the United States; half of the world's known supply of coal is found in North America. In fact, these untapped deposits of coal represent nearly 90 percent of the U. S. fossil-fuel reserves. The actual potential of the U. S. coal reserve is limited to some extent, however, because only a portion can be extracted with present technology or without using more energy in the extraction and transportation than could be realized from the coal obtained. Some of the coal reserves lie deep in the earth, in remote areas, or in deposits contaminated with other substances. In almost all cases, the "net energy" that the deposits represent after extraction costs are subtracted is very low.

In addition, both the extraction and the use of coal produce serious pollution problems. Much of the coal remaining in the United States lies in surface deposits that would have to be "strip-mined." Strip mining, called by its supporters "surface mining," involves the use of gigantic machines that strip away the topsoil so that veins of coal can be easily reached. This kind of mining not only constitutes a visual insult to the earth but also pollutes rivers, lakes, and even underground water systems by releasing poisonous chemical substances. Recent state and federal legislation restricts such devastation somewhat by requiring reclamation of strip-mined land. Stricter regulations are expected in the future, but it seems unlikely that reclamation would ever be able to completely restore strip-mined land to its original condition. Underground mining certainly does less damage to the environment than strip mining, but it has its own problems. Among them are respiratory ailments such as black lung disease, which is common among miners.

The pollution problems caused by the use of coal are even more serious than those caused by petroleum use. The burning of coal to generate electricity produces large amounts of sulfur dioxide and other noxious gases, as well as quantities of coal ash. Technology is now available to remove most of this material so that it will no longer be

A strip mine in the eastern United States

dumped into the atmosphere, but many electric power companies claim that the removal is too costly, cutting down on profits and causing increases in electric rates.

Natural Gas

Natural gas is geologically related to crude oil, and the presence of one is a good indication of the presence of the other. According to F. Donald Hart, president of the American Gas Association, natural gas provides more than 31 percent of the primary energy used in the United States. Space heating, water heating, cooking, and various commercial processes depend on natural gas. It is a cleaner fuel than oil, but it is more difficult to transport. Like oil, natural gas is in relatively short supply and can be only a short-term energy solution. In recent years, the known reserves of natural gas in the United States have dropped steadily; in 1974, the total supply of natural gas was as low as it was in 1952, but 60 percent more gas was being used. Experts predict that the nation's supply of natural gas will not last much longer than the reserves of petroleum.

NUCLEAR ENERGY

Nuclear energy, once believed to be the ultimate answer to any shortages of energy, has instead become a major disappointment for many people. Creating energy by means of the *fission*, or splitting, of uranium atoms, nuclear reactors now provide about 4 percent of the energy used in the United States. But there are many problems associated with the production of nuclear power. Not only is there a very real possibility of the irreversible contamination of the environment by radioactivity, but

These huge tanks encased in concrete will provide storage for radioactive wastes from a nuclear power plant in South Carolina.

uranium is also in short supply and will not last many more decades. In addition, no adequate solution has been found to the problem of the disposal of radioactive wastes from nuclear power plants. Finally, the threat of political hijacking of nuclear fuels (which are commonly shipped by trucks, trains, and planes) or the seizing of the reactors themselves holds frightening possibilities.

Some nuclear scientists hope that development of a new kind of nuclear reactor known as a "breeder reactor" will solve some of their problems. Breeder reactors would operate more efficiently than present-day reactors and would produce nuclear fuel as a byproduct of their operations. Experimental breeder reactors have been built in the United States and several other countries, but so far none is suitable for large-scale commercial use. The development of a *fusion* reactor, which would produce energy by combining atoms of the plentiful element deuterium, has been even less successful.

In 1979, an incident occurred that made it uncertain

whether nuclear energy had any kind of future in the United States. On April 4, a serious accident took place at the Three Mile Island nuclear power plant near Harrisburg, Pennsylvania. As a result of the accident, radioactive steam was discharged from the plant and a nuclear reactor came close to the dangerous condition known as a "meltdown." Although the meltdown was avoided and the reactor shut down without further damage, the incident proved that the risks of using nuclear energy were very real.

GEOTHERMAL ENERGY

The heat contained within the earth has long been looked at as a source of energy and recently has attracted increased attention. Geothermal energy, in the form of steam from hot springs and other underground sources, can be used to operate turbines that generate electricity. It is now considered a potential source of a limited but significant amount of power.

There are geothermal fields located around the world—in Iceland, Italy, Japan, the Soviet Union, Mexico, New Zealand, and the United States. Some of these sources of power are already being tapped. The world's first geothermal power plant began operating in 1904 in Larderello, Italy, and is now producing 380,000 kilowatts of power. A geothermal plant located near San Francisco has operated since 1960, producing 192,000 kilowatts. Other areas in the United States, classified as steam fields by the U. S. Geological Survey, are potential sources of geothermal energy. In 1970 the passage of the Geothermal Act allowed the leasing of public lands for the development of known geothermal areas.

A geothermal power plant near San Francisco

One drawback in the use of geothermal energy has been the corrosion of equipment by the minerals contained in the hot geothermal waters, the control of which increases the costs of power plants. In some areas of the world, air pollution resulting from the use of geothermal energy could be serious since various types of noxious gases such as hydrogen sulfide are often byproducts of thermal wells. The disposal of waste water that contains large amounts of minerals and the control of noise and ground disturbances are all problems that must be solved if geothermal plants are to be acceptable.

OTHER SOURCES OF ENERGY

As we have seen, most of the energy used in the world today comes from sources that are limited in supply and that cause various kinds of pollution. There are some currently used forms of energy production, however, that appear to be infinite in supply and generally clean and safe to use. The two that come to mind most readily are water power and wind power—both indirect forms of solar power.

Water Power

Water power owes its existence to the energy of the sun, which is responsible for the *hydrological cycle*—the endless recycling of the earth's waters. The heat of the sun causes evaporation of water from oceans, lakes, and rivers. This water vapor rises and forms clouds that eventually return moisture to the earth as precipitation. Because of the hydrological cycle, the waters of the earth continuously change form and move from one place to another. This

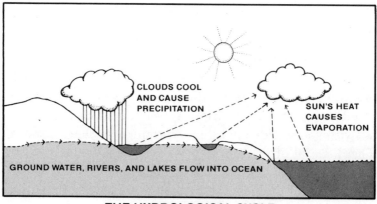

CLOUDS COOL
AND CAUSE
PRECIPITATION

SUN'S HEAT
CAUSES
EVAPORATION

GROUND WATER, RIVERS, AND LAKES FLOW INTO OCEAN

THE HYDROLOGICAL CYCLE

A modern hydroelectric plant on the Feather River in California

natural cycle turns salty ocean water into pure water and provides moisture for growing plants. It also makes possible the use of flowing water as a source of power. According to M. King Hubbert, retired United States Geological Survey expert on solar energy, water power represents the "largest concentration of solar power that is produced by any natural process."

Human use of this form of solar power is very ancient, dating from before the time of the Roman Empire. In more recent periods, the flowing waters of rivers and waterfalls have been used to turn waterwheels that have powered machines in textile and flour mills, in lumbering operations, and in light manufacturing. The modern use of water power has concentrated on the production of electricity, using the force of water flowing over dams or waterfalls to provide power for electric generators. In the United States today, there are five hydroelectric plants with power-generating capacities exceeding one million kilowatts each, enough to supply a city the size of Cleveland with energy.

Although such hydroelectric plants are not a source of pollution, they do cause some environmental problems. In some cases, scenic valleys and canyons have had to be flooded in order to build hydroelectric dams in the United States. Many people feel that the loss of such beautiful country, when there is so little wilderness left, is not worth the additional electricity generated. Another potential drawback of hydroelectric power is caused by the sediment carried by the waters of dammed rivers or streams. This sediment is deposited in the reservoirs behind the dams and will eventually fill them up, preventing the operation of the hydroelectric plants. To date, the problem of sedimentation has not been solved.

Hydroelectric power is most often produced by damming rivers and streams, but it can also be produced by harnessing the energy in rising and falling ocean tides. *Tidal power plants* are set up near bays or coastal basins where tides are particularly strong. Dams are used to trap the rising water at high tide and to hold it back when the tide begins to fall. At low tide, when the difference between the two water levels is the greatest, the water behind the dam is released. Release of the water at the proper time produces the maximum amount of power to drive electric turbines.

The Rance River tidal electric plant

The world's first successful tidal electric plant, completed in 1966, was erected at the mouth of the Rance River in northern France. It is capable of producing 320 million kilowatts of electricity. In the United States, a major tidal-power project, planned for an area in northern Maine, has been under discussion since the 1920s. Called the Passamaquoddy Tidal Power Project, it would make use of the tides in the Bay of Fundy, known to be among the strongest in the world, to produce electricity. Although the Passamaquoddy Tidal Power Project was enthusiastically supported by President Franklin D. Roosevelt during the 1930s and '40s, funds were never allocated for its construction. In the 1970s, however, the United States Congress expressed renewed interest in the project.

While tidal power could produce only a small percentage of the world's total energy requirements, individual tidal projects could supply almost all the needs of areas nearby. Tidal power is a clean and safe source of energy. It produces no pollution and causes minimum disturbance to the environment.

Wind Power

Wind is another indirect form of solar power. The earth's winds are produced by the heat of the sun, which warms the surface of the earth unevenly. The air over hot areas expands and rises, and the air from cooler areas flows in to take its place. This movement of air in the earth's atmosphere produces the strong prevailing winds that occur over large sections of the earth and the local winds that influence limited areas.

People have used the energy of the wind for many centuries. In the early history of the human race, wind was used to move ships and to turn windmills that pumped water for irrigation. During the Middle Ages, windmills and water wheels were the primary sources of power. For centuries, windmills have dotted the landscape in The Netherlands. In 1800, some 11,000 windmills were being used in that country; in 1954, over 1,000 of them were still in operation, 400 being used to drain the watery land and

600 to mill grain. Windmills have also been put to work in the United States, particularly during the 18th and 19th century. According to a recent survey conducted by New Mexico State University at Las Cruces, about 15,000 windmills are in working order today in the United States. The survey reported that another 70,000 could be restored.

Windmills have been used as a source of energy for many centuries.

Wind power is pollution free, relatively inexpensive, and available locally almost anywhere in the world. The energy of the wind is enormous, and it is mostly wasted on the earth today. Yet this ancient, traditional source of power is still used in isolated spots, and it will undoubtedly become significant again. In a later chapter, we shall see that there are some plans for its redevelopment.

Wind power and water power are both manifestations of the enormous influence of the sun on the earth. The next chapter of this book will describe some of the other ways in which solar energy affects life on earth.

2

The Sun
as a Source
of Energy

The view of ancient sun worshipers that the sun is life-giving and all-powerful is closer to the truth than many people realize in today's modern industrial world. Too often, we unthinkingly name the gas or electric utility company as the source of energy in our everyday lives. In fact, as we have seen, the sun is the ultimate source of all the energy stored in the fossil fuels, as well as the energy produced by the earth's wind and water. What is this fiery ball in the sky that is so important to life on earth?

In astronomical terms, the sun is a yellow dwarf star located on the edge of the galaxy known as the Milky Way. It is composed of extremely hot swirling gases that are bound together by a gravity 28 times that of the earth. While the sun is the controlling force in our solar system, it is only one of the billions of stars in the universe.

The sun is an enormously large body compared to the earth. Its diameter is approximately 864,000 miles (1,382,400 kilometers), which is 109 times the diameter of

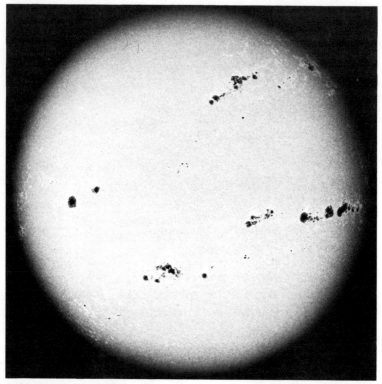

A photograph of the sun showing sunspots, patches on the sun's surface where the temperature is lower than in the surrounding area

the earth. The mass of the sun is also many times greater than that of our planet. This can best be seen from a simple comparison: if the earth weighed one ounce, the sun would weigh 10 tons! Approximately 93 million miles (148,800,000 kilometers) separate the earth from the sun; light from the sun takes about 8 minutes and 20 seconds to travel that great distance. (Light from Proxima Centauri, the star next nearest the earth, takes more than four years to reach us.)

By studying the spectrum of light from the sun, scientists can tell much about the elements of which it is composed. Estimates are that the sun is made up of about 70 percent hydrogen and 28 percent helium. The other 2 percent of the sun's mass consists of traces of many other elements, including carbon, nitrogen, oxygen, magnesium, and silicon.

Another way of learning about the sun is by studying its different regions. The surface of the sun, which gives off energy in the form of heat and light, is called the *photosphere* (a word of Greek origin that means "sphere of light"). The photosphere is the golden ball we see in the sky. Above or around the photosphere is the *chromosphere* (sphere of color), which is made up of fiery red clouds of gases that sometimes shoot out as far as a million miles (1,600,000 kilometers). Around these two spheres is a more diffuse region called the *corona*, another area of gases that can extend more than nine million miles (14,400,000 kilometers) from the sun. The chromosphere and the corona are visible naturally only during total solar eclipses, but knowledge of them has greatly increased since the invention of the coronagraph, a device that can produce an artificial eclipse in an observatory.

The *interior* of the sun is a region of extremely dense gases and high temperatures. The temperature at the sun's core is estimated to be nearly 27 million degrees Fahrenheit (15 million degrees Celsius). It is in this innermost region that the sun's heat and light are produced by means of thermonuclear reactions. These complicated reactions occur when the extreme heat and pressure in the sun's core cause hydrogen to be changed into helium. In the process, part of the sun's matter is transformed into energy and released in the form of heat and light. The reaction that takes place in the sun is basically the same as the one that occurs at the explosion of a hydrogen bomb, but the energy given off by the sun is equal to millions of hydrogen bomb explosions every minute. The earth is protected from the immense heat and the radiation given off in this continuous explosion by its distance from the sun and by its sheltering atmosphere. This protection is not absolute, however, as anyone who has ever gotten a severe sunburn has discovered.

The total amount of energy given off by the sun is inconceivably large. It has been estimated that the sun radiates the equivalent of the heat produced by 400 billion trillion tons of burning coal every hour. The upper atmosphere of the earth receives only one two-billionth of the

Solar prominences are arches of gas shooting out from the sun's surface. This prominence, photographed at an observatory in California, is 272,000 miles (435,000 kilometers) high.

sun's total energy, but even this fractional amount is equivalent to the energy represented by 126 billion horsepower.

All the solar energy that reaches the earth's upper atmosphere does not penetrate to the surface. In an article in *Scientific American* (September 1971), M. King Hubbert estimated that about 30 percent of solar energy reaching the earth bounces directly back into space, about 47 percent is absorbed by the earth and converted into heat, about 23 percent is spent driving the hydrological cycle, and less than 1 percent furnishes all the energy responsible for the movement of air and of ocean waters. Most amazing, only 0.03 percent of the solar energy reaching the earth is captured by plants and used in the process of photosynthesis. Yet it is this tiny percentage that provides all the food energy for living creatures on earth, not to mention all the stored fossil-fuel energy discussed earlier.

As these figures indicate, most of the enormous energy that the earth receives from the sun is not used by humans directly. Of course, we do benefit from the sun's natural heating of the earth, but we could gain so much more. The potential for additional benefits from the sun's energy is almost unlimited. And since scientists estimate that the sun will continue to exist for billions of years and to produce almost boundless energy, we would do well to take a good look at its potential. But before we consider the future of solar energy, let us take a quick look at the past.

A carving showing the Egyptian pharaoh Ikhnaton
and his wife worshipping the sun god Aton

3

Human Attitudes toward the Sun

For thousands of years, the inhabitants of the earth have venerated the sun, the "glorious lamp of heaven." There is no way of knowing exactly when worship—or even awareness—of the sun began. Many historians believe, however, that appreciation of the sun corresponded to the human race's first efforts at cultivating the soil and making things grow. Scientists think that such efforts date back 7,000 to 10,000 years to what is called the Neolithic, or New Stone, Age. During this period, humans discovered that seeds planted in the ground would eventually sprout and produce grain. The early farmers soon realized that most things planted where the sun shone all day thrived, whereas those in the shade usually did poorly. From this realization must have come a gradual awareness of the sun as a mysterious force that was somehow the source of food and thus the source of life.

Another gift of the sun was heat. It did not take long for prehistoric people to realize that the rising of the sun in the morning brought warmth and comfort. In the evening, however, when the sun fell from the sky, it often became cold, uncomfortable, and too dark to see threatening danger. Confronted with occurrences that were so very basic to their existence, it is not surprising that early humans began to worship the life-giving power of the sun. They knew that the sun was somehow supporting them and enabling them to exist.

SUN WORSHIP

About 7,000 years ago, around 5000 B.C., the Egyptian people of the Nile Valley developed a system of agriculture based on the growing of dates, figs, olives, onions, and grapes. The Egyptians made great strides in agriculture, using complex farming techniques based on such sophisticated concepts as preparing the land for planting and irrigating crops. Such a highly agricultural society was bound to recognize the importance of the sun in daily life. The early Egyptians were the first people able to calculate the solar year as 365 days. The sun also played an important role in Egyptian religion. Among the many gods the early Egyptians worshiped was the sun god Ra, who they believed was rowed across the sky each day in a boat. During the 1300s B.C., the Egyptian pharoah Ikhnaton preached a new religion dedicated to a single god, the sun god Aton, who was represented in images as a disk sending out rays. In his "Hymn to the Aton," Ikhnaton wrote:

At daybreak, when thou arisest on the horizon,
When thou shinest as the Aton by day,
Thou drivest away the darkness and givest thy rays. . . .
Whatever flies and alights,
They live when thou hast risen for them. . . .

The fish in the river dart before thy face;
Thy rays are in the midst of the great green sea. . . .
Thy rays suckle every meadow.
When thou risest, they live, they grow for thee.
Thou makest the seasons in order to rear all that thou
    has made,
The water to cool them
And the heat that they may taste thee. . . .

The Trundholm sun chariot

Sun worship in the ancient world was not confined to the Nile Valley. During the Bronze Age, people in what is now Scandinavia, for example, also considered the sun an object of adoration and reverence. Many rock and cave engravings and other artifacts survive today to give evidence of that fact. The famous Trundholm sun chariot, made in about the 13th century A.D., shows a sun disk being transported in a chariot. Many engravings found in the Scandinavian area from this period portray "disk-

At the summer solstice (around June 21), the sun rises directly over this stone marker at Stonehenge. The ancient monument in England may have been used as an observatory or as a site for sun worship.

people," round humanoid figures associated with a sun image.

Sun worship may also have existed in ancient Britain. On the Salisbury Plain, near what is now London, a circular setting of enormous vertical stone slabs known as Stonehenge was constructed sometime between 1800 and 1400 B.C. The meaning of this mysterious arrangement is not definitely known, but since it casts certain shadow patterns at the time of the summer solstice and since it resembles a sundial in some ways, many archaeologists believe that Stonehenge was connected to ancient rituals of sun worship.

In the classic worlds of Greece and Rome, the sun played an important role in legends and myths. According to Greek and Roman mythology, the god Helios, or Sol, represented the sun and was driven across the sky from east to west in a chariot each day. The Greeks called Helios "the yellow haired" and depicted him as a beautiful young man with long flowing locks and a beard that was stylized in drawings to represent rays of the sun.

On the American continents, several major cultures engaged in forms of sun worship. During the 14th and 15th centuries A.D., the Aztec civilization in the Valley of Mexico had a religion centered around the god Huitzilopochtli, a warrior and sun figure who died at the setting of the sun every night and was born anew as the sun rose in the morning. At dawn Huitzilopochtli fought a battle with the moon and the stars, using a shaft of light as his weapon. The Aztecs believed that in order to thank the sun god and to make certain that he would keep up his work, they had to make human sacrifices. Thousands of lives were offered each year at the sun temple in the Aztec capital, Tenochtitlan. Most of the victims were prisoners taken during the many wars that the Aztecs fought with neighboring tribes.

The Incas of Peru also worshiped the sun at about the same period, though they sacrificed animals rather than people, and then not in great numbers. The Incas built beautiful temples in honor of the sun, the most extravagant of which was in their capital city, Cuzco. The finest gold was used to cover the surfaces of the temple that received direct sunshine, and the brilliance of the building on a clear sunny day was breathtaking.

Many North American Indian tribes worshiped the sun, but the Natchez Indians of what is now the southeastern United States probably had the most elaborate form of sun worship. The Natchez chief, who was known as the Great Sun, functioned as a priest-ruler, serving as an intermediary between the people and the sun. In the Natchez temples, a perpetual fire was kept burning in honor of the sun.

While sun worshipers have existed during many historical periods and in widely different cultures, human veneration of the sun has shown some remarkable similarities, as we have seen. In early civilizations, the common denominator of sun worship was the realization that the sun was a source of heat, light, and life. Eventually, people discovered that they could enlist the power of the sun to further serve the needs of the earth.

The Aztec Sun Stone has an image of the sun god at its center, surrounded by symbols representing the days of the week and important events in Aztec history.

4

Early Experiments in Using Solar Energy

No one knows exactly when humans first used the sun for purposes other than growing food. Historians believe, however, that during prehistoric times, people probably learned that they could expose pools of salt water to the sun's rays and eventually obtain both salt crystals and drinkable water. Interestingly enough, the principle of such solar stills remains basically the same today.

The first written account of the direct use of the sun comes from Greek history. According to some sources, the Greek scientist and engineer Archimedes used the sun to win a major battle in the year 212 B.C. When the Roman fleet began to attack Archimedes' native city, Syracuse, the scientist supposedly stationed men bearing hundreds of mirrors along the tops of the city walls. The men positioned the mirrors so that they reflected sunlight on the sails of the attacking Roman ships, and the sails burst into flames! With the aid of a good wind, the fire spread from ship to ship, and the entire Roman fleet was destroyed.

The account of Archimedes' feat makes an exciting story, but historians are not sure whether this event actually took place. They do know that the Incas, in their worship of the sun, used silver mirrors to light fires for ceremonies in the temple at Cuzco. In Europe at about the same time, people were using magnifying glasses to concentrate the sun's rays. By the 17th century, "burning glasses," as they were often called, were considered part of any scientist's equipment. In 1695, two Italian scientists named Averoni and Targioni used a large glass and the sun's rays to melt a diamond belonging to a certain duke. History does not record whether the nobleman was reimbursed for the gem's loss.

Earlier in the 17th century, a French scientist, Solomon de Caux, had constructed a machine activated by the sun that was able to pump water. Another French scientist, Antoine Lavoisier, experimented with using sunlight to produce extremely high temperatures. He focused the sunlight with a large lens made up of two glass disks; the space between the disks was filled with *wine* to improve the lens's refraction. In 1774, Lavoisier used a 52-inch (130-centimeter) main lens and an 8-inch (20-centimeter) secondary lens to achieve a temperature of 3,118 Fahrenheit (1,750 degrees Celsius) in his solar furnace. In this furnace, the scientist could melt metals and conduct chemical experiments.

In 1866, the emperor Napoleon III engaged the services of still another French scientist, Auguste Mouchat, to devise ways of utilizing the sun's energy. Mouchat designed the first known solar food cooker. His hot-house method made use of a box with a glass cover that allowed the sun's rays to pass through, trapping the heat to cook

Antoine Lavoisier's solar furnace

the food. Mouchat also developed a sun-powered steam engine that could be used to pump water.

Solar-energy pioneers were at work in other parts of the world during the 19th century. In the United States, the Swedish-born inventor John Ericsson built a hot-air engine powered by the sun in 1826. Ericsson attached a solar reflector to the engine in such a way that when the sun shone, the engine would be kept running. The reflector could be pivoted so that it was always aimed at the sun. During the next 50 years, Ericsson worked at perfecting similar solar engines.

Near the end of the 19th century, a French inventor, Abel Pifre, built a solar engine that could be put to

One of John Ericsson's solar engines

commercial use. Pifre used his engine to run a printing press that printed a newspaper aptly called the *Sun Journal (le Journal Soleil)*. Solar energy was used in other commercial ventures during the early years of the 20th century. In 1901, on an ostrich farm in Pasadena, California, a solar motor was developed that was capable of pumping water at the rate of 1,400 gallons (5,320 liters) per minute at its highest efficiency. In 1912, an American engineer named Frank Shulman set up a solar irrigation project in Egypt. The solar plant worked successfully, pumping water from the Nile River. But it was not very popular with the local people, whose hand labor it had replaced. The resulting labor disputes, as well as the outbreak of World War I, caused the plant to be shut down after only a few months of operation.

These are only a few examples of commercial solar ventures; the list could go on and on. Historical events, however, eventually discouraged the continuing development of solar energy in the early part of the 20th century. The supplies of coal, oil, and natural gas that had been discovered in the preceding century finally were found to be so plentiful and inexpensive to use that there seemed to be no point in developing other sources of energy. Solar inventions came to be classified as oddities and gadgets, curiosities that had little or no practical application. This situation continued until the 1950s, when some far-sighted scientists and engineers realized the limited nature of the world's conventional fuel supplies and again began looking to the sun for solutions to the energy problem. Today, solar technology is becoming more and more attractive to the nations of the earth as fuel prices rise, supplies diminish, pollution increases, and world political conditions become less and less stable.

The main mirror of the Odeillo solar furnace in France

5

Solar Energy Today

What can the sun do for people? A great deal, as we shall see. The sun's energy can be used to heat and cool houses, schools, hospitals, and commercial and industrial buildings. It can be used to pump water and to distill it, to run solar furnaces, to cook food, and to dry agricultural products such as corn, tobacco, and grapes. Solar energy can evaporate ocean water to obtain salt, operate communications and navigation systems, power space vehicles, and generate electricity. In order to put the sun to these varied uses, a special solar technology is needed. Through much experimentation, several basic methods of using the sun have been developed that offer not only hope for a bright solar future but also solar power for present use.

SOLAR COLLECTORS

The basic requirement of almost all forms of solar technology is some means of gathering together the scattered energy of the sun's rays. The simplest way of doing this is to use some kind of *solar collector,* a device that collects the sun's radiation and turns it into heat. The many varieties of solar collectors can generally be divided into two categories: *flat-plate collectors* and *focusing collectors.* A third kind of solar collector, the *parabolic trough collector,* combines some of the features of both the flat-plate and the focusing devices.

Flat-Plate Collectors

A simple flat-plate solar collector consists of five essential parts:

1. An absorber plate that intercepts, collects, and absorbs solar radiation
2. One or more panes of glass or plastic located about an inch above the absorber plate
3. A series of pipes usually attached to the back of the absorber plate
4. A 3- or 4-inch thickness of insulation beneath the absorber plate
5. A rectangular frame enclosing the whole device.

The absorber plate is usually made out of metal—steel, aluminum, or copper—and coated with a black material such as carbon black to increase the absorption of solar radiation. The glass panes allow the solar radiation to penetrate to the absorber plate, at the same time preventing heat from escaping. (This is the well-known "greenhouse effect," the same principle that causes the extreme heat inside a car parked in a sunny spot with the windows shut.) The heat trapped in the collector is absorbed by water or air flowing through the pipes behind the absorber plate. The water or air then carries the heat into whatever kind of system the collector is providing energy for: usually a water- or space-heating system in a building.

Flat-plate collectors are usually installed on the roofs of buildings, facing south so that they receive as much exposure to the sun as possible. When collectors are used to

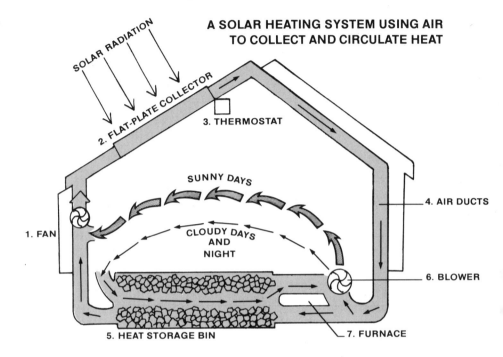

A SOLAR HEATING SYSTEM USING AIR TO COLLECT AND CIRCULATE HEAT

SOLAR RADIATION

2. FLAT-PLATE COLLECTOR

3. THERMOSTAT

SUNNY DAYS

CLOUDY DAYS AND NIGHT

1. FAN

4. AIR DUCTS

6. BLOWER

5. HEAT STORAGE BIN

7. FURNACE

1. FAN circulates air through heating system.
2. FLAT-PLATE COLLECTOR absorbs heat from the sun. Air flowing through collector picks up heat.
3. THERMOSTAT controls system, turning fan on to bring cool air to collector and off when heat is not needed or sun is not shining.
4. AIR DUCTS carry heated air to heat storage bin or to blower.
5. HEAT STORAGE BIN contains rocks that absorb heat to be used at night and on cloudy days.
6. BLOWER blows heated air into house directly from collector or from heat storage bin.
7. FURNACE provides supplementary heat when needed.

ON SUNNY DAYS, warm air from the collector is blown directly into the house.
ON CLOUDY DAYS AND AT NIGHT, warm air from storage bin is circulated through the house.

supply hot water only, the equipment needed is very simple, little more than a water tank and conventional plumbing. In a solar-powered space-heating system, a series of ducts or pipes within the building is required to distribute the heat with the aid of pumps or blowers. (In many cases, conventional heating systems in existing buildings can be adapted for this purpose.) Heat is stored for future use in insulated tanks or, in the case of a system using air, in bins full of rocks.

Flat-plate collectors can also be used for cooling. A typical cooling system works on basically the same principle as the gas refrigerator, using a coolant such as ammonia to absorb heat. Although the technology of solar cooling is not as simple nor as advanced as that of solar heating, future developments should make it possible to heat *and* cool buildings efficiently using the sun's energy.

Flat-plate collectors are in use all over the world today, especially in countries like Japan, Australia, and Israel. So far, most of the collectors are being used to heat water. A growing number of homes and office buildings, however, are being heated and cooled at least partially by solar collection systems. According to a recent study, there were 10,000 houses in the United States using solar energy for space and water heating in 1980. Many of these homes are located in the sunny Southwest, where solar energy can supply almost 100 percent of their heating needs. In less sunny parts of the country, supplementary heating systems are needed for extended periods of cloudy, cold weather.

In addition to residential use, solar energy has also been put to work in public institutions. The United States government, in cooperation with private industries, has in-

Flat-plate collectors are used to provide heat for this public school in Osseo, Minnesota.

stalled solar collectors at public schools in four different locations—Osseo, Minnesota; Warrenton, Virginia; Baltimore, Maryland; and Dorchester, Massachusetts. The collectors provide supplemental heating, supply hot water, and even keep swimming pools warm. The general areas chosen for the experiment represent challenging locations for the utilization of solar energy—especially Minnesota and Massachusetts, where cold winters are common. The state of Minnesota is also the location of another interesting solar-energy experiment, an underground bookstore on the Minneapolis campus of the University of Minnesota that is partially heated and cooled by solar collectors. In Ventura, California, where sunshine is a little more dependable, a 254-unit apartment complex has been built that uses solar energy for all its space and water heating. Tenants of the apartments pay nothing for the energy supplied by the solar-collector system and less than $5 a month for the natural gas that is used for heating on cloudy days.

A huge focusing collector produces high temperatures in the world's largest solar furnace.

Focusing Collectors

Flat-plate collectors use the sun's energy to produce temperatures high enough to heat homes and supplies of water. The highest temperature that can be achieved by flat-plate collectors is about 400 degrees Fahrenheit (about 204 degrees Celsius), whereas focusing collectors can produce temperatures up to 6,350 degrees Fahrenheit (3,500 degrees Celsius). This type of collector uses mirrors to focus the sun's rays on a central point. In most focusing collectors, the primary mirror is curved in the shape of a *parabola*. (A parabola is a curve that directs rays hitting it

48

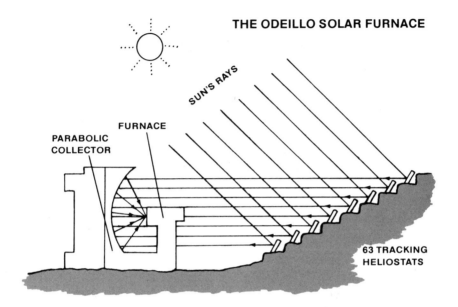

THE ODEILLO SOLAR FURNACE

SUN'S RAYS

FURNACE

PARABOLIC
COLLECTOR

63 TRACKING
HELIOSTATS

to a certain specific location with great accuracy.) These parabolic mirrors, made of glass or spun aluminum, must be produced with great precision in order for the collector to create the highest possible temperatures.

Focusing collectors are used primarily by industries that require high temperatures for various processes. Today, the most common application for the focusing collector is the *solar furnace.* The world's largest solar furnace is the Odeillo furnace, located at a French research facility in the Pyrenees Mountains. Designed by Dr. Felix Trombe and built in the late 1960s, the Odeillo furnace includes a parabolic collector that is 133 feet (40 meters) high and 165 feet (49.5 meters) wide. Also part of the system is a series of smaller flat mirrors, called *heliostats,* which are located on a hillside in front of the main mirror.

The heliostats are constructed so that they turn and tilt automatically to follow the sun as it moves across the sky. Throughout the day, the heliostats reflect the sun's rays on the huge parabolic collector, which is made up of small mirrors accurately curved to focus at one point—a small building that is the furnace itself. Objects within the furnace are subjected to temperatures as high as 5,975 Fahrenheit (about 3,300 degrees Celsius)—hot enough to melt steel plates or to create artificial gems.

In addition to the Odeillo furnace, there are a number of smaller solar furnaces in the world today that also produce high temperatures. In White Sands, New Mexico, the U. S. Army operates a solar furnace that can achieve temperatures of 5,000 degrees Fahrenheit (2,760 Celsius). Second only to the Odeillo furnace in size, the White Sands facility is used to study the effects of thermal radiation on missile components. Corvair, an aerospace company located in San Diego, California, has utilized a solar furnace to determine the heat capacities and the thermal shock characteristics of ceramics. Other industries and institutions have put solar furnaces to work in Japan, Russia, Algeria, and Argentina.

Parabolic Trough Collectors

The parabolic trough collector combines some of the features of the flat-plate and focusing collectors. It can achieve temperatures higher than those of flat-plate collectors but not as high as temperatures produced by focusing collectors. Developed by the University of Minnesota and Honeywell, Inc., the parabolic trough collector uses a curved, trough-like device to concentrate solar radiation on a "heat pipe" located at the focus point of the collector.

The pipe is enclosed in a glass tube, which, like the glass covering of a flat-plate collector, prevents thermal energy from being radiated back into space. Within the pipe is a liquid, usually water, which is heated by the sun's rays to a temperature of 575 degrees Fahrenheit (about 302 degrees Celsius).

Like flat-plate collectors, parabolic trough collectors can be used to supply energy for heating and cooling. Honeywell's office building in Minneapolis, Minnesota, is

Parabolic trough collectors on the roof of the Honeywell parking ramp in Minneapolis, Minnesota. The collectors provide 35 percent of the energy needed to heat a nearby office building and 50 percent of the energy needed for air-conditioning.

partially heated and cooled by a group of 252 trough collectors located on top of a parking ramp near the building. The heat produced by parabolic trough collectors can also be used to create steam that can provide power for electric turbines. (The exciting possibilities of this aspect of solar technology will be discussed in the following chapter.)

SOLAR STILLS

When we talk about shortages in the world today, we usually think of fuel and food. Yet many scientists predict that there will be severe shortages of fresh water by the 21st century, primarily because of pollution. Of course, in some areas of the world there has always been a shortage of drinkable water, and sometimes a complete lack of it. For the inhabitants of these areas, and increasingly for people everywhere, solar distillation of water, particularly ocean salt water, is going to become very important. By means of solar distillation, the sun can produce fresh water for drinking, irrigation, and industrial use. As a by-product of the distillation process, the sun can also produce salt!

The principle of solar distillation is very simple. In a solar still, salt water is collected in shallow ditches or tanks that are covered by slanted pieces of glass or plastic. When the heat from the sun is absorbed on the bottom of the tank, the water becomes warm enough to evaporate. The water vapor rises until it reaches the glass or plastic cover, which is cooler than the bottom of the tank. This difference in temperature causes the vapor to condense (turn to liquid) on the underside of the covering. The distilled water then runs down the cover and into collecting

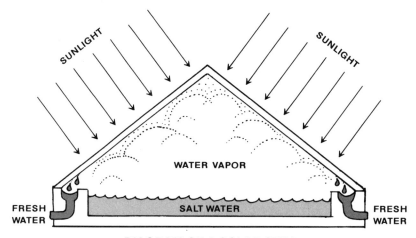

DIAGRAM OF A SOLAR STILL

troughs. The water is fresh and pure since the salt, which evaporates at a higher temperature than water, has been left behind in the tank.

The solar still described here is a very simple one; there are many more sophisticated models. All work on the same basic principle, which is very similar to the principle of the flat-plate solar collector, except that water, rather than heat, is the desired end product.

The world's first large-scale solar distillation plant was built in 1874 in the middle of the Atacama Desert at Las Salinas, Chile. There was a valuable potash mine at Las Salinas, but, as the Spanish name of the location suggests, the only water available was too saline—too salty—to drink. In order to supply water for the miners and for the donkeys used to transport the potash, a solar still covering an area of 51,200 square feet (4,608 square meters) was constructed. The still produced 6,000 gallons (22,800 liters) of drinkable water every day until 1914, when fresh water was finally piped into the desert.

The solar still at Coober Pedy, Australia

Solar stills are functioning today, particularly in the warm and sunny parts of the world. In the community of Coober Pedy, Australia, well water is too salty to be used. Drinkable water had to be hauled in by truck until 1966, when a solar still was constructed. This still produces almost 2,895 gallons (11,000 liters) of pure water each day from the saline well water. Solar energy is also being used in the production of salt. The Leslie Salt Company of California uses the natural process of evaporation to produce about a million tons of salt yearly at their 29,000 acres of salt flats north of San Francisco.

Although solar distillation is not extensively used at present, it may eventually prove to be the only answer to the water shortage in many remote areas of the world. And the search continues for more economical ways of using the sun to make sea water drinkable for people everywhere.

SOLAR COOKERS

Can you bake a cherry pie using the sun? Absolutely! One of the most interesting applications of solar energy has been in the kitchen—or rather in moving the kitchen to the backyard. Numerous stoves and enclosed ovens have been designed that can cook hot dogs and roasts, boil water, and bake cakes. Basically these stoves follow the patterns of flat-plate and focusing collectors; some of them operate by trapping heat under glass, and others concentrate the sun's rays on the food. While such solar cookers have not yet been mass-produced, they are being

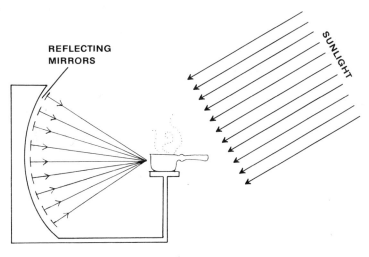

REFLECTING MIRRORS

SUNLIGHT

DIAGRAM OF A SOLAR COOKER

introduced in many tropical countries as a clean, cheap, and portable source of heat for preparing food. An obvious drawback is that, unlike large collectors, existing cookers provide no heat storage for a rainy day. It is hoped that further research will solve this problem.

It now appears that the first wide-scale application of solar energy in the United States will be in the heating and cooling of homes and eventually of larger buildings. Honeywell, Inc., a company deeply involved in solar-energy research, estimates that by 1981 home solar heating and cooling units will be available on the market at reasonable prices. Since heating and cooling accounts for 25 percent of energy use in the United States today, utilization of solar energy in this area alone would result in a significant easing of the energy shortage and a reduction of pollution created by conventional energy sources. In the next chapter, we shall look at some of the future applications of solar energy for which technologies are now being developed.

6

The Solar Future

As we have seen, fairly simple solar energy devices can be used to provide space heating and hot water directly to individual buildings and to cook food. In order for solar energy to be employed on a large scale, however, it must be turned into a more usable form, one capable of operating all the complicated machinery of modern life. One of the most universal and versatile forms of energy is electricity, and scientists have developed a number of methods that may be used in the future to produce electricity from sunlight.

SOLAR-THERMAL CONVERSION

Several of the methods of making electricity employ techniques similar to those used to heat houses or to power solar furnaces. They involve the use of solar collectors of various kinds to concentrate the sun's energy and to produce heat. Heat is not the end product of the system, however, but instead is used to raise the temperature of water to the boiling point. The steam produced by the boiling water can then be used to operate conventional steam turbines like the ones found in thousands of electrical power plants throughout the United States today. The

A drawing of the National Solar Power Facility as proposed by Aden and Marjorie Meinel. In this illustration, the area between the rows of flat-plate collectors is being used as grazing land for cattle.

turbines in turn provide the power for generators that produce electricity. This basic process of using the sun's energy to generate electricity is usually referred to as *solar-thermal conversion.*

There have been several exciting plans for solar-thermal conversion projects proposed in recent years. One of the most interesting is that devised by a team of scientists led by Aden and Marjorie Meinel of the University of Arizona. The Meinels have proposed an enormous solar collecting and generating plant to be constructed in the desert area of the southwestern United States. They believe that *all* the nation's electricity could be produced at this vast plant, which they have named the National Solar Power Facility.

The National Solar Power Facility would be erected on 5,000 square miles (13,000 square kilometers) of unused desert land in the adjoining states of Arizona, California, and Nevada. It would include row after row of flat-plate collectors that would tilt automatically to receive the most sunlight possible during different times of the day and year. The heat absorbed by the collectors would be carried by a system of pipes to a central power plant, where it would be used to heat water and make the steam necessary to operate steam-turbine generators.

Area to be covered by the
National Solar Power Facility

The Meinels have developed several new methods of making their energy-collecting system as efficient as possible. One involves the use of a special black coating on the collectors that will aid in the absorption of heat. Another new idea is the use of liquid sodium instead of water to transport the heat from the collectors to the central power plant. Liquid sodium, a form of salt, would hold more heat than would water and would permit more efficient storage of heat for use during the sunless hours of the night.

Once the heat from the collectors is transported to the central power plant, an adequate supply of water must be available for the production of steam. Water is not abundant in the desert Southwest, but the Meinels propose the building of large aqueducts that would bring in seawater from the Gulf of California. The seawater would be converted to steam, used to power the turbines, and then condensed back into a liquid. By this time, however, it would have been transformed into fresh water, some of which could be used to irrigate the arid desert.

The Meinels' plan for the National Solar Power Facility has received much attention in recent years, and not all of it has been favorable. Some scientists and engineers oppose the idea of having one location as the source of all the nation's power, for a variety of reasons. For one thing, they point out that the cost of electricity would increase with the distance over which it is transported from its source to its point of use. Another problem would be the grave consequences of disaster at a national power-producing facility. The destruction of the nation's single source of electric power by either a natural calamity or one caused by humans would immobilize the entire United States. Many scientists who are opposed to the idea of the National Solar Power Facility suggest instead a number of smaller facilities, strategically located around the country.

Another possibility would be the construction of solar power-generating plants that would serve individual communities. G. E. Brandvold of Sandia Laboratories in New Mexico has proposed a total energy community project, which would involve "on-site" collection of solar energy by a central group of flat-plate collectors. Part of the solar energy would be used to produce steam and to generate electricity. The rest of the collected heat would be distributed within the community and used for water heating, space heating, and air conditioning.

There are several advantages in such on-site generation of electricity. One would be the reduced cost of transmission of electric power from the plant to the home. Another would be the possibility of using the waste heat from the steam turbines for purposes of heating and cooling. This heat is carried away from the turbines by the water used to cool and condense the steam. In conventional fuel electric plants, the heated water is often dumped back into rivers and lakes, causing thermal pollution. In a community solar electric plant, however, it could be put to good use in nearby homes and schools.

The question of on-site versus centralized solar power production is one that comes up often when the future of solar power is discussed. Both methods have advantages and disadvantages, and a combination of the two will probably prove to be the most practical in the long run.

Both the on-site and the centralized plans that we have considered up to this point involve the use of flat-plate collectors. (Parabolic trough collectors could also be used effectively in some of these schemes.) Other solar electric projects depend on focusing collectors like the kinds used in solar furnaces. One such plan, now in the experimental stage, is the solar power tower. A power tower facility consists essentially of a tall tower upon which is mounted a water boiler. Surrounding the tower is a field of heliostats, which are designed to tilt and rotate in an automatic tracking of the sun. The heliostats collect and reflect the sun's radiation onto the boiler, and the steam produced in the boiler as a result is piped to a conventional steam-turbine generator at the base of the tower. The steam can also be transferred to thermal storage for use at night when the sun is not shining.

An artist's concept of a solar power tower

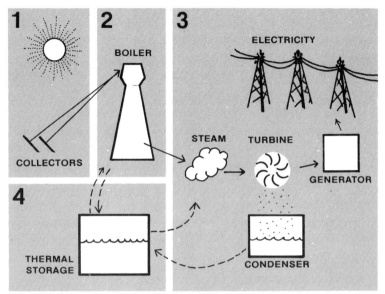

A SIMPLIFIED PLAN OF A SOLAR POWER TOWER FACILITY
A power-tower facility has four basic parts, or subsystems: 1. The collector subsystem, which collects solar radiation and transmits it; 2. The receiver subsystem, which receives the radiation and uses it to make steam; 3. The electrical power generation subsystem, which produces electricity by means of a steam-turbine generator; 4. The thermal storage subsystem, which stores the heat from solar radiation for later use.

The federal government has given contracts to four large American firms for developing and testing a solar power tower plant. In 1979, a pilot plant began operation in Albuquerque, New Mexico, using 9,550 mirrors to focus the sun's rays on a 200-foot-tall tower. Tests at this plant may confirm the opinion of experts who believe that the power tower method is one of the most economical means of producing electricity from solar radiation. It can be used in both on-site and centralized power production.

OCEAN-THERMAL ENERGY CONVERSION

Solar-thermal conversion is one method of using the sun's energy to make electricity. There are other methods even more exotic. One of the most exotic is *ocean-thermal* energy conversion, a means of using the difference of temperature in ocean water to produce electricity.

In the tropical oceans of the world, the sun-heated surface water is often 30 to 35 degrees Fahrenheit (17 to 19 degrees Celsius) warmer than water 1,500 feet (450 meters) below the surface. As early as 1881, a French physicist, Jacques d'Arsonval, found that a heat engine could be operated by using this temperature differential. Another French scientist, Georges Claude, designed and built a successful ocean thermal power plant off the coast of Cuba in 1929. A second plant was built off the coast of Africa in 1956. Today a number of experimental projects are underway using ocean thermal principles.

A typical ocean thermal plant, like the prototype designed by Lockheed Missiles and Space Company of California, consists of a floating cylinder-shaped device that takes in warm water from the surface of the ocean and cold water from the ocean depths. The warm water passes through boilers containing liquid propane or ammonia. When these volatile fluids are heated, they turn into a vapor that can then be used to power electric turbine generators located beneath the boilers. After the vapor has done its job, it is cooled and condensed by cold water drawn in at the bottom of the power plant. Returned to its liquid form, the propane or ammonia is ready to be recycled to the boilers and vaporized once again. The cold water used in the condensation process is discharged back into the ocean a few degrees warmer than before.

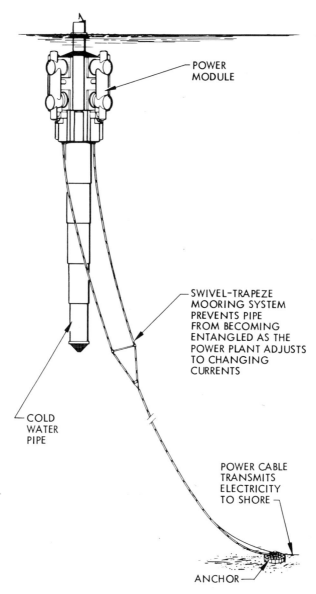

POWER
MODULE

SWIVEL-TRAPEZE
MOORING SYSTEM
PREVENTS PIPE
FROM BECOMING
ENTANGLED AS THE
POWER PLANT ADJUSTS
TO CHANGING
CURRENTS

COLD
WATER
PIPE

POWER CABLE
TRANSMITS
ELECTRICITY
TO SHORE

ANCHOR

A diagram of the ocean-thermal power plant developed by
Lockheed Missiles and Space Company

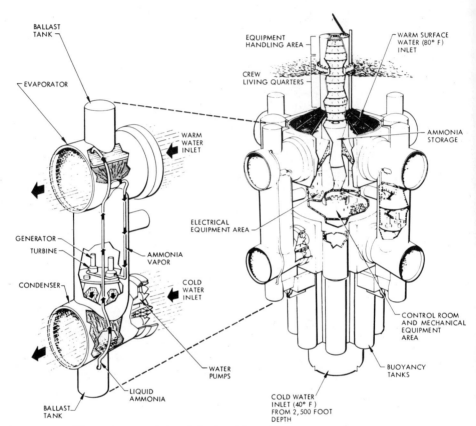

The power module of the Lockheed ocean-thermal plant

In the United States, ocean thermal plants would be placed off the coast of Florida in the warm waters of the Gulf Stream. The electricity produced by the plants would be sent ashore by means of underwater cables. There would be no need for energy storage, since ocean temperatures remain fairly constant even at night and on cloudy days. And of course there would be no fuel costs involved in the production of ocean thermal electricity, except for the initial cost of the fluid used to power the turbine generator. Although this amazing method of using the sun's energy is still in the experimental stage, its potential for the future is enormous.

ELECTRICITY FROM THE WIND

Another means of producing electricity involves the use of windmills. We have already seen that using the energy of the wind is a very old concept. But now this idea has put on a new face, and wind power is taking its place in the "big league" of alternative energy sources. Not only have individual windmills on farms and homesteads regained respect, but engineers and scientists are also looking at ways in which large "wind-farms" may be developed to produce electricity on a large scale.

People who live in such areas of the United States as the Great Plains and the regions around the Great Lakes or the Maine shore know what the word *wind* means. Such areas are subjected to almost continuous battering from the wind; in fact, people who live in these regions consider the fierce wind an enemy. Some scientists have suggested that large groups of windmills could be constructed in such places to harness this now wasted energy.

When windmills are used in the production of electricity, the whirling blades themselves serve as a turbine, providing the power needed to operate an electric generator. Since the wind does not blow at a constant rate, however, the power supply is not a steady one. Modern experimental windmills, or wind turbines, have been built in shapes designed to get as much use as possible from varying wind speeds and directions. Research has also

Modern wind turbines look very different from traditional windmills.

been done on methods of storing the energy produced by windmills for use during windless periods. One method would employ the process of *electrolysis,* a means of using electricity to extract hydrogen from ordinary water. (A water molecule is made up of two atoms of hydrogen and one of oxygen; when an electric current is passed through water containing a substance known as an electrolyte, the water molecules break up, separating into hydrogen and oxygen.) Hydrogen either in gaseous or liquid form can be used as a fuel for many purposes. The hydrogen produced by means of a electricity-generating windmill could be stored and burned later to generate more electricity during periods when the wind is not blowing.

There are some problems to be solved before wind power can be used as an energy source on a large scale. But continued research may help to realize the great potential of this form of solar energy.

SOLAR CELLS

All the solar-energy projects discussed so far use mechanical devices to produce electricity. They employ various forms of solar energy to operate turbines that in turn operate electricity-producing generators. There is, however, a much more direct method of producing electricity from the sun's energy. It involves the use of *solar cells,* remarkable devices that generate electricity directly when exposed to sunlight.

Solar cells, also known by their scientific name *photovoltaic cells,* are thin discs or rectangles usually made up of two layers of silicon. Mixed into the silicon layers are very small amounts of substances such as arsenic, alumi-

A photograph of the Skylab space station in orbit around the earth. One of Skylab's solar-cell wing panels was lost when the space station was launched.

num, or boron. When sunlight strikes the silicon atoms in the presence of these substances, it sets up an electric current that flows between the two layers of silicon. The current produced by solar cells can be used to provide power for electric lights, to run machinery, and to do thousands of other jobs in today's industrial world.

Solar cells sound like something dreamed up by a science fiction writer, but they have actually been in limited use for almost 20 years. The American and Russian space programs have used panels made up of thousands of solar cells to provide electric power for many kinds of space vehicles. For example, the Skylab space station launched by the United States in 1973 was equipped with several large wing panels of solar cells. In 1973 and 1974, Skylab spent a total of 171 days in orbit around the earth. During this time, nine astronauts lived on board and carried out scientific experiments with the aid of electric power supplied by solar cells.

Producing large numbers of solar cells such as those used on Skylab is extremely expensive.

Solar cells have been indispensable to the space program, but until recently, they have been so expensive to manufacture that they have not been considered practical for any large-scale production. Many companies, however, have been attempting to develop cheaper and more efficient methods of making solar cells. Between 1975 and 1979, the cost of producing a single watt of electricity from solar cells was reduced from $22 to around $9. The Department of Energy believes that the cost can be reduced to 30¢ per watt by 1990. If this goal is achieved, then in the future panels of solar cells may be used to provide energy for homes, offices, schools, and even electric automobiles and airplanes.

An artist's concept of a Solar Satellite Power Station

One revolutionary plan for the large-scale use of solar cells has been proposed by Dr. Peter Glaser of the Arthur D. Little Company. Dr. Glaser has suggested building enormous solar-cell collectors in space. The collectors, called Solar Satellite Power Stations, would be placed in orbit around the earth in such a way that they would always remain above the same spot on the earth's surface. Each collector would have several wing panels seven and a half miles long! The electricity produced by the mass of solar cells on the panels would be transferred to another orbiting satellite by means of large cables. There the electricity would be converted to microwaves (the form of radio waves used in radar and in microwave ovens), which would be beamed to a gigantic receiving antenna on earth.

The microwaves would then be transformed into usable electric power and transmitted to private and industrial users of electricity.

There would be many advantages to such orbiting solar power stations. Positioned more than 20,000 miles (32,000 kilometers) above the earth's surface, the solar-cell collectors would receive sunlight unobscured by clouds and other forms of interference in the earth's atmosphere. Two collectors placed in orbit far enough apart would allow the solar power system to receive and transmit energy 24 hours a day; while one collector was cut off from the sun by the earth's shadow, the other would be in full sunlight. Of course, constructing such an orbiting solar power system would be an enormous and expensive undertaking. Dr. Glaser and other scientists believe, however, that the United States has the necessary technical knowledge and that the task would be simpler and less expensive than that of putting humans on the moon. Time will tell whether this method of using solar energy will prove to be practical.

RESEARCH IN SOLAR ENERGY

Solar cell conversion is one of the four primary methods of electric power generation selected by the United States government for research and development in its exploratory solar energy program. The other three methods are solar-thermal conversion, ocean-thermal conversion, and wind energy. Another area that the government will explore in the future is *bioconversion,* the use of plant materials to produce electric power. As we have seen, green plants convert the energy of the sun into food energy during the process of photosynthesis. The stored energy of plants that decayed millions of years ago has been transformed into the fossil fuels that we now use as sources of power. Bioconversion would provide a means of speeding up the natural process by which the energy stored in plants is converted to fuel.

One method of bioconversion involves growing large amounts of plant material specifically for use as fuel. For example, plants such as eucalyptus, sugarcane, and water hyacinth grow quickly and provide an abundant source of *biomass,* a word used to describe vegetation to be converted to energy. Scientists envision large "energy plantations" where such plants would be grown, then cut down,

Water hyacinths are fast-growing plants suitable for use in the process of bioconversion.

dried, and burned to provide steam for electric turbines located nearby. Other plans call for treating biomass plants with chemicals and converting them into methane and other gases that could be used as fuels. The fast-growing seaweed kelp is a plant well suited to this method of bioconversion.

Other research projects are attempting to find additional ways of using solar energy to save fuel. At the University of Florida's Solar Energy and Conversion Laboratory, studies are being made on the application of solar energy to the treatment of sewage. Today many sewage plants must use heating fuels to operate the digesters, devices that process the sewage. Solar heating of digesters is being tested in Florida, using plastic sheets glued together to form a kind of pillow or air mattress. The sheets serve as a trap for solar energy, somewhat like a flat-plate collector, and provide enough heat to keep the digesters operating properly.

These are just a few of the exciting ideas for the use of solar energy that may become reality in the future. Many experts involved in solar research believe that there are no limits to the potential of sun power except the limits imposed by the human imagination.

7

Solar Energy: Pro and Con

Why do so many people—scientists and nonscientists alike—believe that solar energy may be the answer to the energy problems of the modern world? Some of the reasons are obvious; solar energy has a number of advantages over the traditional energy sources on which we now depend. Probably the most important advantage is that solar energy is virtually inexhaustible. There is no evidence that the sun will cease to exist for a billion years, so for our purposes its energy supply is infinite. No shortages will occur, no reserves will be depleted, and no wells will run dry.

Another significant advantage of solar energy is that it is clean. The use of the sun will not cause smog, will not pollute drinking water, will not require strip mining of land or flooding of canyons. No refining process with all its dirty byproducts will be needed, nor will it be necessary to transport fuels that can spill, contaminating oceans and rivers. There will be no problem of disposing of dangerous waste products from solar installations. In fact, the lack of any polluted discharge from solar electric plants will make it possible to build them in the centers of towns.

And, wonder of wonders, solar energy is free! Of course, this does not mean that using it will be free. The initial investment in equipment will, in fact, be considerable. But the fact remains that the "fuel" itself is free and equally available to all nations (although those located near the equator will have a slight advantage). No longer will some countries be "rich" in energy, while others are at their mercy because of a lack of oil or coal.

Moreover, in contrast to other energy sources, solar energy will be available at a predictable rate. We know that the sun will shine at least a certain number of days each year and that the earth will receive a certain amount of solar energy. Long-term planning will be much simpler with a solar-based system; it will not have to be done on the basis of the latest drillings for oil in Alaska or the political situation in the Middle East.

These are some of the many advantages of solar energy. Of course, there are also some disadvantages involved in utilizing the power of the sun. Let's take a look at a few of the most significant problems that must be overcome.

Isn't the sun's light too diffuse and too weak to supply a large amount of energy? Wouldn't too much land be needed for collectors?

Enough solar energy reaches the United States in just 20 minutes to supply power to the entire country for a full year. This sunlight *is* spread very thin, however, and it must be trapped and concentrated, as we have seen, if we are to use it efficiently.

Collectors to concentrate the sun's energy would take

A model of a solar power farm using flat-plate collectors

up quite a bit of space. In his book *Energy Crises in Perspective,* John Fisher has estimated that if the total energy supply of the United States were to depend on flat-plate solar collectors, about 50,000 square miles (130,000 square kilometers) of land would be needed to set up the equipment. This might sound like a lot of land (certainly more than the Meinels' estimate of 5,000 square miles), but actually it is only about 3 percent of the area now devoted to farms and only about .013 percent of the nation's total land area. Of course, development of other forms of collection systems, such as the use of the tower collectors described earlier, might significantly reduce the amount of land needed.

Isn't the weather too unreliable, too unpredictable, for solar energy to be depended upon? What happens at night? Or during cloudy periods?

Actually, from a meteorologist's point of view, weather is fairly predictable. Meteorologists know what the average number of sunlit days will be in any part of the United States during the year, and the numbers are surprisingly high, as the map shows. It would probably be easier to predict how many hours of sunshine any American city will have five years from today than to predict how many

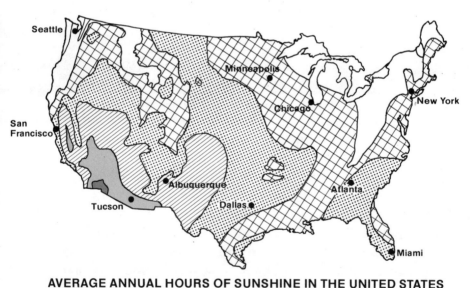

AVERAGE ANNUAL HOURS OF SUNSHINE IN THE UNITED STATES

| 4,000 | 3,600 | 3,200 | 2,800 | 2,400 | 2,000 or less |

oil wells will have dried up by that time.

Of course, there will always be periods when the sun does not shine—at night and during rainy or cloudy weather. For this reason, the development of solar technology will have to include efficient methods of storage. In addition, simple back-up systems using other forms of energy will be included in solar-heated and -cooled homes and offices. These systems will be used during cloudy periods too prolonged to be dealt with by stored solar energy. Eventually, the most practical long-range method of storing solar energy may prove to be the technique of making hydrogen from solar-produced electricity and storing it for later use.

Won't air pollution adversely affect the use of solar energy?

Air pollution *does* cause problems in the use of solar energy; there is no question that in a city like Los Angeles today, a solar collector would operate at less than its peak of efficiency. In most areas of the country, however, including much of the industrialized East Coast, air pollution has come under state and federal controls and is being abated. Improvements continue daily. And as solar energy begins to be used for heating, cooling, and for other purposes, the amount of pollutants in the air will be decreased proportionally. The problem of air pollution cannot be overlooked, but it is not a major drawback to the use of solar energy. Since flat-plate solar collectors can operate on diffuse light (not just direct sunlight), they can still function under less than ideal conditions. As long as efforts to limit air pollution continue, the solar future will not be threatened.

Won't the use of solar energy cause thermal pollution?

Thermal pollution occurs when heat used to create electric power or to operate an industrial process spills out into the environment, usually into rivers and lakes. Thermal pollution can kill or injure marine life and otherwise affect the ecology of the surrounding area. It is a legitimate concern in any discussion of solar energy since high temperatures *will* be reached in some solar-powered processes. This problem has not been completely solved, and continued research will be required. As we have seen, one possible way of handling thermal pollution may be to put it to work. Since solar power plants are so clean, they can be located very near or within cities so that waste heat can be utilized on the spot by industries and homes.

Won't solar collectors and other forms of solar equipment cause "visual" pollution? Won't they be ugly and clumsy-looking?

This is a valid criticism of some of the present solar equipment, but there is no real reason why solar collectors, which are basically simple in design, cannot be handsome additions to the landscape. The American Institute of Architects has expressed great interest in working with solar-energy equipment, and architects of the future will make every effort to design aesthetically pleasing solar buildings. Some solar equipment is already quite pleasing to the eye, as anyone who has seen photographs of the Odeillo solar furnace can testify. Finally, solar energy will improve our visual environment by replacing some existing eyesores; we will never see a solar collector or a solar power plant spoiling the view by pouring filthy soot out into the air.

The Odeillo solar furnace

Won't the introduction of solar energy cause a lot of legal problems?

Yes, it will. The lawyers and the courts will be kept busy dealing with the variety of problems that will arise. For example, the thousands of building codes that exist in the United States today will have to be altered. A new concept, that of "sun rights," will have to be applied to all construction. No longer will someone be able to build a tall building that would prevent the rays of the sun from reaching other buildings nearby. Land-use planning will also be affected by the introduction of solar energy. The potential for chaos cannot be minimized; the task of adapting a nation to this new form of energy will be a challenge to the legislative and judicial systems.

83

Won't solar power also cause difficulties and confusion in the construction industry? And within labor unions?

Yes, there will be complicated problems of jurisdiction to iron out here as well. For example, who would be responsible for the construction of a rooftop flat-plate solar collector in a new house: the carpenter who is building the house; the plumber who is responsible for all pipes and water flow; or someone from the heating and air conditioning trade? Would a collector built by a home-owner—a nonprofessional—be approved by city building inspectors? At present there are very strict codes governing the construction industry in the United States, and solar energy would require rewriting many of them. In the same manner, labor unions have strict rules about who performs what construction services and how much workers are paid for their work. The coming of solar energy will make drastic changes in all these areas.

Won't many workers in the oil and gas industries be put out of work if solar energy becomes successful?

The new solar industry should employ enough people to compensate for jobs lost in other industries. In the early years of the 20th century, many workers who were employed in the carriage and wagon industries and those responsible for providing horses protested the coming of the "horseless carriage"—the automobile. The new automobile industry, however, employed many more workers than those whose jobs disappeared. The same thing may happen with the development of solar energy. Even if solar power does not become a large-scale industry, the oil and gas industry will eventually die out as resources are used up. Then the jobs it provided will no longer exist.

Won't the use of solar energy be extremely expensive?

This is an important and complicated question. The initial cost of developing technology for solar energy will be high indeed. When considering the cost of solar energy, however, we must think in terms of a long period of time. Developing a solar-energy technology is the kind of investment that will eventually pay for itself, as compared to investments in systems that use fossil fuels. For example, if you spend $3,000 on a solar collector for your home but save $300 in fuel bills every year by using the sun's energy,

A solar collector made up of glass tubes. Equipping homes with such collectors will not be cheap, but it will be an investment that will eventually pay for itself.

then in 10 years you will have paid for the collector. And during the 11th year and all subsequent years, the use of your heating and cooling system will be free.

This is a simplification, of course, since you would undoubtedly have maintenance costs for the collector and other pieces of equipment. The principle, however, is still valid. And the economic benefits of using solar energy can only increase with time. As oil and gas become more and more scarce, their cost will continue to rise. But the economics of supply and demand will never apply to the sun.

Solar energy is a good investment not only for individuals but also for nations. Although national governments will have to commit large amounts of money to solar technology initially, in the long run this investment will be cheaper than trying to pay for Middle Eastern oil. And when the world's supplies of oil and other fossil fuels are eventually depleted, a nation's investment in solar technology will really prove its worth.

Won't the cost of solar systems be too much of a personal financial burden on the individuals who first install them?

People who install solar systems may need to be given special financial help. One way of easing the burden is by offering tax credits; the federal government has already set up a program that allows a maximum $4,000 tax credit on the installation of solar heating equipment. The United States Department of Energy has also suggested a number of other ways in which the government can encourage consumer acceptance of solar energy and aid those who install solar systems. Some of its suggestions are: 1. Providing aid

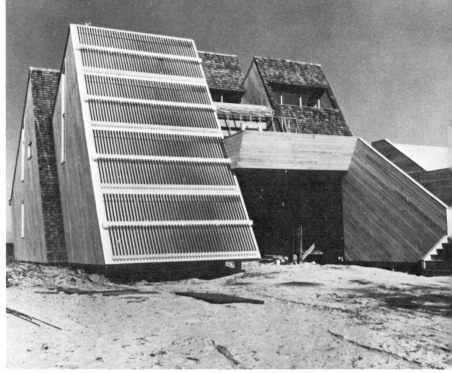
This unusual house on New York's Long Island is heated by solar collectors.

to banks and other financial institutions that offer mortgages for new solar homes or for older homes retrofitted with solar equipment; 2. Offering incentives to insurance companies that would provide insurance for such homes; 3. Providing mortgage or insurance funds directly to buyers on a temporary basis; 4. Encouraging owners of rental dwellings to install solar systems by providing tax breaks; 5. Providing incentives to the construction industry to develop solar-heated and -cooled buildings.

In addition to these plans by the federal government, a number of states have passed legislation designed to encourage and aid solar development. States such as Arizona, Colorado, Illinois, Maryland, New Hampshire, New Mexico, North Dakota, Oregon, South Dakota, and Texas have taken this step, often using tax breaks as a means of aiding individuals who invest in solar equipment.

Most of the plans for the future use of solar energy seem to concentrate on home heating and cooling and on the production of electricity. But what about transportation? Can solar power play any role in this area of modern life?

This question is a valid one, and important in any discussion of solar energy. Even if all the electricity needed in the United States were supplied by solar energy and all buildings were heated and cooled by sun power, about 25 percent of the nation's energy needs would still be unfilled. This 25 percent is the portion required for transportation—moving people and goods from place to place. Solar energy can play a role in providing power for transportation. If solar cells eventually prove their worth, electric automobiles, trains, and planes could be powered by panels of solar cells. Many scientists believe, however, that the most practical method of using solar energy in transportation would be in the production of hydrogen by means of electrolysis. Hydrogen in its gaseous form can be used as a fuel in much the same way that gasoline is now used. Hydrogen is an efficient fuel, but it does have some disadvantages; it requires a large amount of space for storage, and it can be dangerous to handle. If methods can be found to counteract these difficulties, then solar-produced hydrogen may become the transportation fuel of the future.

8

Planning for the Solar Future

As recently as the 1960s, proposals for using the sun to produce energy on a large scale were laughed at by many and filed in the wastebasket. Words such as "exotic" and "impractical" were heard whenever harnessing the power of the sun was suggested. But by 1979, when gasoline cost more than one dollar a gallon and drivers waited in long lines to fill their tanks, the laughter began to die out. Then people began to take a second and more serious look at the idea of solar energy.

Today many nations throughout the world have embarked on some exciting and ambitious solar programs. Japan is one of these nations. Among the world's most highly industrialized societies, Japan must at present import all but 2 percent of its fuel requirements. To achieve more independence, the Japanese government has initiated a billion-dollar program, "Sunshine Project," to encourage the use of solar energy and to insure the integra-

Solar water heaters on the roof of a house in Japan

tion of the solar industry into the nation's economy. As a part of Sunshine Project, the government is financing the development of large-scale centralized electric generation facilities powered by solar energy. It is also promoting the use of solar-produced hydrogen gas as a fuel. Solar water heaters are already being mass produced and widely used in Japan and, in fact, are even exported to other countries.

In France, the government scientific research agency has set up the Institute of Solar Energy to coordinate the work of over 300 scientists and engineers involved in solar

These solar collectors will be used to provide hot water for the apartments of the British Embassy staff in Brazilia, the capital of Brazil.

research. West Germany has committed itself to multimillion dollar solar research and development projects to be initiated during the next three years. The Swiss government is now considering a proposal to install solar heating and cooling systems (called "climate control systems" in Switzerland) throughout the country.

Australia is another nation that has made a serious commitment to the use of solar energy. In certain parts of Australia, solar hot-water heaters are required by law. Legislation is also pending that will make possible large-scale construction of solar stills in areas without drinkable water. The Soviet Union is concentrating on solar power to be used in its vast undeveloped regions. Soviet scientists have already designed solar refrigerators, solar-powered sluice gates for irrigation projects, and solar stills that provide sheep and cattle with drinkable water in desert areas.

A few of the world's developing countries have also begun to put solar energy to use. For example, in the nation of Mauritania, located on the southern fringe of the Sahara Desert, inhabitants of a certain village now have twice as much water available as in the past, thanks to the installation of a solar-operated water system on the roof of the local school house. Because of high costs, however, most developing nations cannot afford to pay for the initial research and construction necessary for large-scale solar energy utilization.

The United States, one of the world's richest nations, is

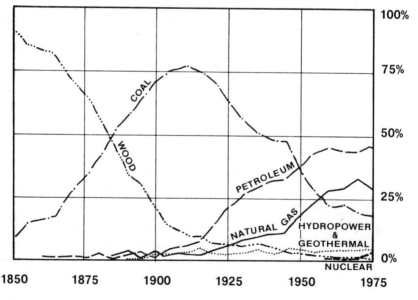

SOURCES OF UNITED STATES ENERGY

of course much better able to afford the cost of solar energy. Moreover, the nation has a very high stake in the solar future. With only 6 percent of the world population, the United States uses 33 percent of the world's total energy production. This large consumption of energy makes possible the high standard of living to which Americans have become accustomed. If the nation is to maintain such a high standard (and many experts believe that this may not be possible under any circumstances), then it must be able to rely on an adequate supply of energy. And the sun seems to be the only energy source that the United States can depend on for future needs.

But how will the use of solar energy be introduced into American life? Traditionally, the introduction of such a new idea or product takes place by way of the "free marketplace." Acceptance and development progresses at whatever pace the public demands, and often the pace is very slow. But solar energy cannot be developed by this slow process if it is to be a workable energy source by the time that other sources have become unreliable. The introduction of solar energy in the United States will have to be a planned venture sponsored by the federal government. It will have to involve all segments of society, including state and local governments, colleges and universities, industries, labor unions, the scientific community, and, of course, the consumer. Such a cooperative effort between government and the free enterprise system is an absolute necessity if the widespread use of solar energy is to occur reasonably soon.

There are some guidelines to look to in achieving this goal. In recent history, the United States government has generated two other major drives to accomplish techno-

The United States government has already put solar energy to use in its military program. This solar furnace at White Sands Missile Range in New Mexico is used to test the effects of thermal radiation on missile components.

logical goals—one to develop nuclear power, and the other to land humans on the moon. In both cases, a deadline was established, no limits were placed on the price tag, and the task was accomplished. S. David Freeman of the Ford Foundation has compared the development of solar energy to these other accomplishments and has said, "I would suggest that Congress establish a national goal to perfect the commercial availability of solar energy for large-scale commercial use within 10 years. By large-scale I mean not only for heating and cooling buildings . . . but also the development of electric power production. Such a goal would be ambitious, but certainly more easily attainable and much less costly than our national decision to go to the moon."

In 1974, the first step toward the achievement of that goal was taken when Congress passed the Solar Energy Research, Development, and Demonstration Act. The purpose of the act was "to pursue a vigorous program of research and resource assessment of solar energy as a major source of our national needs and to provide for the development and demonstration of means to employ solar energy on a commercial basis."

Included in the Solar Energy Research, Development, and Demonstration Act was a provision for the establishment of the Solar Energy Research Institute (SERI), an organization that would be responsible for the assessment of all solar energy sources (wind, hydro, tidal, direct solar, etc.) both nationally and regionally. In 1977, Dr. Paul Rappaport, an expert on solar-cell technology, was appointed director of SERI, and by 1979, the institute had begun operations in its headquarters near Golden, Colorado.

The establishment of the Solar Energy Research Institute represented a significant advance in American plans for the solar future. In the following years, the government took other steps to encourage and support the development of solar energy. The energy plan proposed by President Jimmy Carter in July 1979 called for new solar tax credits and for the establishment of a "solar bank" that would provide low-interest loans for the installation of solar equipment. The president's plan also set a goal for the future: by the year 2000, solar energy in all its forms should be supplying 20 percent of the nation's energy needs. To achieve that goal, the Department of Energy allocated $800 million to solar energy projects in 1980. But during that same period, nuclear projects received $1.72 *billion*. And the same energy plan that set the ambitious goal for solar energy had some less acceptable features. It called for a lowering of air pollution standards so that high-sulfur coal could be used and for a crash program in the development of synthetic fuels from such nonrenewable sources as coal and oil shale. Looking at these proposals, supporters of solar energy wondered whether the federal government was as committed to the solar future as it claimed to be.

Of course, some government reluctance to give wholehearted support to solar energy is no doubt due to genuine caution and care. The introduction of solar energy into American life is not something that can be undertaken without careful planning or without risk. But there are also other forces at work in the energy game. The fossil-fuel and nuclear industries have been active in encouraging the government to favor already developed energy systems in distributing funds.

These industries have also tried to influence public opinion against new forms of energy. One major oil company, for example, recently sponsored a television ad that presented a rather biased look at the use of windmills. "Back in 1915," the ad went, "3,000 windmills helped light up the country of Denmark. America could generate electricity in the same charming way. All we'll have to do is keep wasting our natural stores of energy. When it's all gone, we'll turn on the windmills. A great idea until the wind dies down." The scene then shows the windmill sails stopping and the lights going out, while a voice is heard cursing the darkness in Danish. What the makers of this ad have chosen to ignore is that whether we humans are wasteful or not, our "natural" stores of energy (in other words, the fossil fuels) are eventually going to run out. The ad also makes fun of a very legitimate *and* natural source of energy, the wind, which may in fact offer a partial solution to the world's energy problems.

As this ad suggests, the oil companies, as well as other members of the American energy "establishment," have not been enthusiastic supporters of solar energy. But there are signs that the tide may be turning. Some major oil companies have even become involved in solar-energy

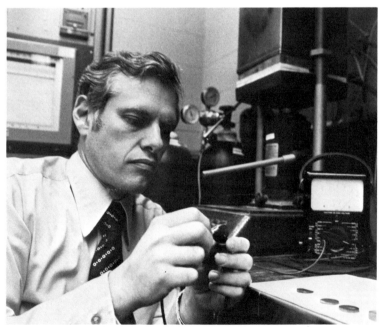

A scientist at the Bell Laboratories tests an experimental solar cell. The Bell System is one of the many companies involved in solar energy research.

research, adding their efforts to those of other organizations and institutions that have begun to prepare for the solar future.

If that future is ever to become a reality, then the efforts of people all over the United States will be required. Ordinary citizens who cannot play a direct role in the development of solar energy can do their part by writing to their representatives in Congress and expressing their commitment to solar power. Such efforts will serve a real purpose, for as environmentalist Egan O'Connor has pointed out, "The question of how soon we will have solar energy on a large scale is first a political matter and only second a technological one." Although much research remains to be done, *the basic technology of solar energy is available now.* All that is lacking are the funds needed to put that technology to work.

Glossary

active solar energy system. A system that uses solar collectors and other special equipment to heat and cool a building. See *passive solar energy system.*

bioconversion. The process of using the solar energy stored in green plants to produce electric power.

biomass. Vegetation used in the process of bioconversion.

electrolysis. A process in which an electric current is passed through a liquid, causing a chemical reaction to take place. When water is subjected to electrolysis, it breaks up into its two elements, hydrogen and oxygen.

flat-plate solar collector. A collector consisting of a flat metal plate painted black and covered with a piece of glass or plastic. Water or air circulating through pipes attached to the back of the plate pick up the heat absorbed by the collector.

focusing solar collector. A collector that uses a system of mirrors to focus the sun's rays on a central point.

fossil fuels. Coal, oil, and natural gas, substances that have been formed by the decay of the remains of ancient plants and animals.

heliostat.   A mirror that turns and tilts automatically by means of clockworks. Used to reflect the sun's rays on a solar collector.

hydrological cycle.   The continuous circulation of the earth's waters, kept in motion by the heat of the sun.

ocean-thermal conversion.   The process of using the differences of temperature in ocean water to produce electric power.

passive solar energy system.   A system of heating and cooling with solar energy that uses the structure of the building itself instead of devices such as collectors.

photovoltaic cell.   See *solar cell.*

solar cell.   A device made of thin layers of silicon or other substances, capable of making electricity directly from sunlight.

solar collector.   A device used to collect and concentrate the rays of the sun.

solar furnace.   A focusing solar collector used to create very high temperatures.

solar-thermal conversion.   The process of using solar energy to generate electricity by means of steam-driven electric turbines; the steam is produced from water heated by solar radiation.

Index

Archimedes, 37
Aton (Egyption sun god), 32–33
Australia, solar energy used in, 46, 54, 91
Aztec civilization, sun worship in, 35

Bay of Fundy, 22
bioconversion, 73–75
biomass, 74–75
Brandvold, G. E., 61
Britain, ancient, sun worship in, 34

Carter, Jimmy, energy program of, 86, 96
centralized solar power production, 58–63, 90
chromosphere, 27
Claude, George, 64
coal, 7, 12–13, 96
Coober Pedy, Australia, solar still at, 54
cooling with solar energy, 46, 56
corona, 27
coronagraph, 27

d'Arsonval, Jacques, 64
de Caux, Solomon, 38
Department of Energy (U.S.), 86–87, 96
deuterium (used as nuclear fuel), 15

Egypt, ancient, sun worship in, 32–33

electricity produced by solar energy, 57, 73; through bioconversion, 73–75; through ocean-thermal conversion, 64–67; through solar cells, 69–73; through solar-thermal conversion, 57–63; through wind power, 67–69
electrolysis, 69, 88
energy crisis, 7–8, 89
energy, world supply of, 7–8, 10, 12, 14, 16, 77–78, 86
Ericsson, John, 39
expense of developing solar energy, 78, 85–87

federal government, U.S., role of, in developing solar energy, 46–47, 63, 73, 93–98
Fisher, John, 78–79
flat-plate collectors, 44, 46–67, 51, 53, 55, 59, 61–62, 75, 80, 84
focusing collectors, 48–50, 55, 62
fossil fuels, 9–14, 74, 86
France, solar energy used in, 90
Freeman, S. David, 94

geothermal energy, 16–17
Glaser, Dr. Peter, 72, 73
Greece, ancient, sun worship in, 34–35

heating with solar energy, 56, 57; space, 44, 45, 46, 47; water, 46, 47, 90, 91
Helios (Greek sun god), 35

heating with solar energy, 56, 57; space, 44, 45, 46, 47; water, 46, 47, 90, 91
Helios (Greek sun god), 35
heliostats, 49–50, 62
Honeywell, Inc., 50, 51–52, 56
Huitzilopochtli (Aztec god), 35
hydroelectric plants, 20
hydrogen as a fuel, 69, 81, 88, 90
hydrological cycle, 18–19, 29
"Hymn to the Aton," 32–33

Ikhnaton (Egyptian pharaoh), 32
Inca civilization, sun worship in, 35, 38
Indians, North American, sun worship among, 35–36
Institute of Solar Energy (France), 90

Japan, solar energy used in, 46, 89–90

Lardello, Italy, geothermal plant at, 16
Las Salinas, Chile, solar still at, 53
Lavoisier, Antoine, 38
legal problems caused by use of solar energy, 83, 84
Lockheed Missiles and Space Company, 64

Mauritania (Africa), solar energy used in, 92
Meindel, Aden and Marjorie, 58–60, 79
microwaves, 72–73
mining, coal, 12–13
Mouchat, Auguste, 38

Natchez Indians, sun worship among, 35–36
National Solar Power Facility, 58–61
natural gas, 7, 14
nuclear energy, 14–16, 95, 96
nuclear reactors, 14–15

ocean-thermal energy conversion, 64–67
Odeillo solar furnace (France), 49–50, 82
offshore oil wells, 10
oil, 7, 9–11, 78
oil embargo, 10
on-site solar power production, 61–63

parabolic mirror, 49
parabolic trough collector, 50–52
Passamaquoddy Tidal Power Project, 21–22
petroleum, 7, 9–11
photosphere, 27
photosynthesis, 29, 74
photovoltaic cells, 69–73
Pifre, Abel, 39–40
pipelines, oil, 11
pollution, 18, 24, 77, 81; caused by use of coal, 12–13; caused by use of geothermal energy, 17; caused by use of nuclear energy, 14, 16; caused by use of oil, 10–11; thermal, 61, 82; "visual," 82

Ra (Egyptian sun god), 32
radioactive wastes, disposal of, 14–15, 16
Rance River tidal power plant (France), 21
Rappaport, Dr. Paul, 95
research in solar energy, 63, 73–75, 95, 97

Scandinavia, ancient, sun worship in, 33
sewage, treatment of, with solar energy, 75
Shulman, Frank, 40
silicon, 69–70
Skylab space station, 70
sodium, liquid, 60
solar cells, 69–73, 88
solar collectors, 43–44, 57, 78, 84; flat-plate, 44, 46–47, 51, 53, 55, 59, 61–62, 75, 79, 81, 84; focusing, 48–50, 55, 62; parabolic trough, 50–52, 62; power tower, 62–63, 79; solar cell, 72–73; temperatures produced by, 48, 50, 51; used to produce electricity, 57, 59, 60, 61–62
solar cookers, 55–56
Solar Energy Research, Development, and Demonstration Act, 95
Solar Energy Research Institute (U.S.), 95
solar furnace, 49–50
solar power tower, 62–63, 79
Solar Satellite Power Stations, 72–73
solar stills, 37, 52–55, 91
solar system, 25
solar-thermal conversion, 57–63
Soviet Union, solar energy used in, 91
space program, solar cells used in, 70–71
Stonehenge, 34
storage, thermal, 46, 56, 60, 63, 81
strip mining, 12–13
sun, 7; composition of, 25, 26–27; energy produced by, 28–29; regions of, 26–27; size of, 25–26; thermonuclear reactions in, 27–28; worship of, 8, 25, 31–36
Sun Journal, 40
"sun rights," 83
sunshine, average yearly amount of, in U.S., 80
"Sunshine Project" (Japan), 89–90
surface mining, 12–13
Switzerland, solar energy used in, 90–91

tax credits for installation of solar equipment, 86–87
thermal pollution, 61, 82
Three Mile Island nuclear plant, accident at, 16
tidal power plants, 20–22
transportation, solar energy used in, 71, 88
Trombe, Dr. Felix, 49
Trundholm sun chariot, 33
turbines: steam, 58, 59, 60, 61, 62, 75; wind, 67, 68

United States: current use of solar energy in, 46–47, 50, 51–52, 56, 63; fuel reserves of, 10, 12, 14; future of solar energy in, 56, 93–98
University of Minnesota, 47, 50
uranium (used as nuclear fuel), 14

water power, 18–22
waterwheels, 19–20
White Sands, New Mexico, solar furnace at, 50
windmills, 22–23, 67–69
wind power, 22–24
wind turbines, 67, 68

About the Authors

Steve J. Gadler has had a long career as a dedicated environmentalist. A retired Air Force colonel, he was appointed to the board of the Minnesota Pollution Control Agency in 1967. Since that time, his appointment has been renewed by three different Minnesota governors. Mr. Gadler has collaborated in writing a series of books dealing with pollution and has published many articles on environmental issues. He is also a registered professional engineer with a degree in electrical engineering and a popular lecturer on subjects ranging from solar energy to food pollution. Mr. Gadler has received citations for his work from organizations such as the Environmental Protection Agency and the Minnesota Environmental Control Citizens Association. He and his wife make their home in St. Paul, Minnesota.

Wendy Wriston Adamson has been deeply involved in the study of the environment both as a writer and as a librarian. The author of two previous books concerned with environmental problems, she has also served as a librarian in the Environmental Library of Minnesota and in the Environmental Conservation Library of the Minneapolis Public Library. Ms. Adamson lives in Minneapolis, Minnesota, with her husband and two children.